A Science Museum
illustrated booklet

FIRE ENGINES

and other fire-fighting appliances

by K. R. Gilbert, M.A., D.I.C.

Her Majesty's Stationery
Office, London 1966

First published 1966
Second impression 1970
SBN 11 290098 4

Cover: Steam fire engine, 1894 (see plate 11)
Frontispiece: A Fireman of the Sun Fire Office 1805, with a Newsham engine in the background (from Costume of Gt. Britain by W. H. Pyne, 1808)
Note: All the items illustrated in this booklet are part of the Science Museum collections

Introduction

This booklet illustrates the history of fire-fighting appliances by means of exhibits to be seen in the collection at the Science Museum.

Fire is a phenomenon in which combustible materials, especially organic substances containing carbon, react chemically with the oxygen of the air to produce heat. Flame arises from the combustion of the volatile liquids and gases evolved and spreads the fire. Fire extinction is generally brought about by depriving the burning substances of oxygen and by cooling them to a temperature below which the reaction is not sustained. Applications of the principle of extinction by smothering are found in the use of carbon dioxide, carbon tetrachloride and powders, which are employed in portable extinguishers for dealing with special types of fire and in the use of foam for fighting oil fires. By far the most important extinguishant, however, by reason of its availability and general effectiveness, is water, It is more effective than any other common substance in absorbing heat and so reducing the temperature of the burning mass, and the steam produced also has a smothering action by reducing the oxygen content of the surrounding atmosphere.

A fire engine is therefore basically a transportable water pump and is in fact referred to by members of the fire service as a *motor pump* or, if it carries an escape, as a *pump escape*. Most fire brigades have at least one *turntable appliance*, which is also fitted with a pump, and several *trailer pumps*. The equipment of a brigade may also include a fire boat, water tenders, hose-laying lorries, a foam tender, an emergency tender, a canteen van, and other vehicles. All these are generically described as appliances. Water is obtained from the public mains, from open water, and for immediate use there may be a *first aid apparatus* on the vehicle consisting of a tank containing 100 gallons of water, a small pump and a hose pipe. The first aid equipment may well be adequate to put out a fire in its early stages. The services of the fire brigade may indeed not be needed at all if a bucket of water or a portable extinguisher containing water or other suitable extinguishing medium is immediately to hand and is used when the fire is still small.

The Romans had a large fire service, the Corps of Vigiles, equipped with pumps, buckets, ladders, fire hooks for pulling off burning roofs, and other tools. The double-cylinder force pumps, the remains of which have been found, are ascribed to Ctesibius of Alexandria, who is believed to have lived in the third century B.C. They were well-constructed of bronze and the delivery pipe was jointed so that it could be pointed in any direction.

With the collapse of the Roman Empire the fire pump was forgotten and only re-invented about the beginning of the seventeenth century. A two-cylinder force pump with swivelling delivery branch, all mounted in a tank on wheels, was described and illustrated in 1612 in a Book of Machines by Heinrich Zeising. This is the earliest known illustration of a fire engine. The addition of an air vessel to maintain pressure in the delivery side of the pump and thus to enable it to deliver water in a continuous stream was probably invented in the mid-seventeenth century by Hautsch of Nuremburg, but did not find its way into English practice for another half century.

A most important contribution to fire fighting was the invention in Holland by van der Heiden of the fire hose, which was introduced into England on the accession of William III. With the use of the delivery hose a close attack on the fire became practicable instead of the less effective long shot from the engine. The wired suction hose also made the supply of the engine with water much easier. In the seventeenth century public provision for fire fighting extended only to the keeping of equipment by the parish authorities. In London after the Great Fire of 1666 the Livery Companies had each to provide one engine, thirty buckets, two squirts, and three ladders. There was no public fire brigade. The *Fire Office* for insurance was, however, soon founded and was followed by other companies in the fire insurance business. (See frontispiece.) The insurance offices, in order to protect properties at risk, set up private fire brigades, which were at rivalry with each other to secure policy holders by competing in the efficiency of the protection provided.

It was not until 1832 that agreement was obtained to amalgamate the insurance brigades in London into a single formation known as the London Fire Engine Establishment and only in 1866 that the brigade came under public control as the Metropolitan Fire Brigade. Fire Brigades are still the responsibility of local authorities of county status and the National Fire Service set up in 1941 was disbanded in 1947 and its components placed under the jurisdiction of the counties and county boroughs. Public fire fighting services in the United Kingdom are thus under the control of some 140 independent authorities. The overall responsibility for securing the efficiency of the service remained with the Secretary of State for Home Affairs and is administered through the Civil Defence and Fire Service Department.

Most of the very early fire engines were small and rested on runners, but wheels were soon fitted to make them easier to transport. By the beginning of the eighteenth century there were several fire engine builders at work, of whom Richard Newsham was the most successful and best known. His largest engine was big enough for twenty men to work. The trend was towards greater size, and by the end of the century the larger engines were horse drawn. The introduction of folding levers or handles made the design of the bigger engines more practical.

Powerful manual engines continued to be built throughout the nineteenth century. The largest manual engine made a century ago by Merryweather and Sons, one of the leading manufacturers, was worked by forty-six men and delivered 220 gallons of water per minute to a height of 150 feet.

The application of steam power to the fire engine came late and the first steamer was built by Braithwaite and Ericsson in 1829, and it still took another thirty years before the steamer was widely adopted. The majority of steam fire engines in use were horse drawn and the steam motor was used only to drive the pump. In 1840 however Paul Hodge produced a self-propelled steam fire engine and such appliances were becoming common by the end of the century. But by this time the internal combustion engine had arrived and rapidly replaced the steam engine both for driving the pump and for traction. The last steam fire engines to be used in London, in 1917, were supplied to the London Fire Brigade in 1902.

The motor fire engine was at first recognisably the descendant of the horse-drawn manual appliance with the firemen seated back to back along each side, but in the 1930's the limousine type of vehicle was introduced to protect the crew from the weather while in transit to a fire.

The comparative lightness of the internal combustion engine led to the introduction of the trailer pump which can be detached from the towing vehicle and manoeuvred into places difficult of access by a fire engine or towed over soft ground by a tractor. A further development was the light-weight pump which can be carried by two men. It is powered by a petrol engine or even by a small gas turbine.

The turntable appliance, which was first produced in Germany in 1902 and adopted in England in 1904, carries an extensible ladder which can be elevated and swung in a complete circle on its turntable. The ladder mechanisms were fully motorised by 1908. This appliance is used not only for rescue work, but for attacking fires in tall buildings by means of the monitor situated at the top of the ladder.

The fire fighting appliances discussed are all transportable pumps or pressurised containers which are brought into action when an outbreak is noticed or in response to an alarm. Automatic fire alarms usually act by the closing of an electric circuit by a switch actuated by a bimetallic strip which responds to a rapid rise in temperature. Other fire detectors are sensitive to smoke or to the presence of ionisation in the air, which is produced by flame.

It remains to mention the automatic fire protection afforded by a sprinkler installation. A network of pipes connected with the water supply is installed in or just under the ceilings of a building and provided with a sprinkler outlet for every hundred square feet. The sprinkler is a valve that opens suddenly and completely when a certain temperature is reached. The early types introduced a century ago depend for their action on the collapse of a fusible soldered link under the influence of heat. The valve of the modern sprinkler is released when a quartzoid bulb nearly filled with a liquid, which expands on heating, shatters at the operating temperature. The flow of water in the system also leads to the opening of a valve at the control point and causes an alarm bell to ring.

1 Hand pumps

The syringe or fire squirt was used in Roman times for fire fighting and continued in use until the nineteenth century. The one illustrated (front) was made about 1750 and was one of three kept ready for use in the Parish Church of St. Dionis-Backchurch, London. The barrel, nozzle, and two side handles are brass castings; the piston is also of brass, packed with hemp, and fitted with a wooden piston rod. Such squirts were kept supplied with water brought in sewn leather buckets.

The addition of an air vessel and valves to the syringe converted it into an efficient hand pump (right). Water was discharged on the down-stroke through a valve into the space between the cylinder and the outer casing. Some of it flowed through the delivery pipe, but the excess water rose above the opening, thus compressing the air in the casing; and this maintained the flow on the up-stroke.

The pump was supplied with the covered pail, which was kept filled with water ready for use. It was manufactured by Shand, Mason, & Co. from 1848 as the *London Brigade* pump.

The stirrup pump (left) is a form of double-acting pump made in large numbers for fighting the fires caused by incendiary bombs in the 1939-45 war. On the up-stroke the piston valve closes and the foot valve opens. Water above the piston is forced through the discharge pipe and water enters through the foot valve. On the down-stroke the foot valve closes and the piston valve opens. Water enters the space above the piston and the discharge continues as the water in the barrel is displaced by the plunger rod.

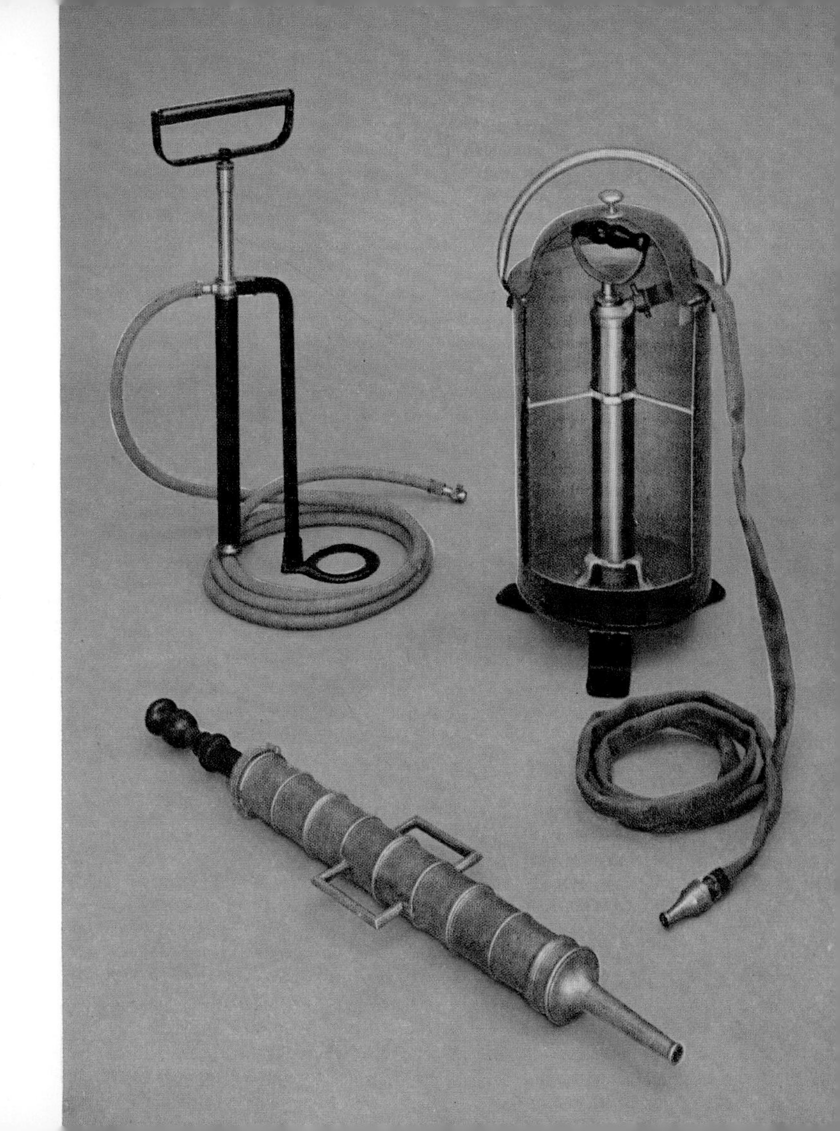

2 Manual fire engine, c. 1680

This model, which dates from about 1680, may have been made by a 17th century fire-engine maker for demonstration to his customers, as the provision of full-size handles out of scale with the model seems to indicate.

It consists of two vertical pumps in a metal cistern mounted on a sled. There is no suction inlet and the cistern has to be filled with water brought in buckets. When the handles are worked up and down the pumps alternately force water into a large copper vessel placed between them. As the water level in the vessel rises above the outlet pipe going to the nozzle, the air is compressed and ejects the water in a continuous stream: without the air vessel the jet would be intermittent. The delivery pipe has two swivel joints in different planes, so that the nozzle may be pointed in any direction.

A print of Keeling's engine, 1678, is reproduced below.

3 The first use of hose, 1673

The earliest book on fire engines was written by Jan van der Heiden and his son Jan, and published in 1690. Van der Heiden, who lived from 1637 to 1712, was both a distinguished artist and Superintendent of the Amsterdam Fire Brigade. He introduced the use of the delivery hose, made of leather, an invention which made the fire engine much more effective than it had been with the swivelling goose-neck delivery jet, by enabling fires to be attacked at close quarters.

The cistern of the engine was connected by flexible hose with a canvas trough in a wooden stand which was sited on a canal bank and kept filled with water by buckets. This obviated the use of a long chain of buckets. Van der Heiden later introduced the wired suction hose.

He was the illustrator as well as the author of his book, which is written in Dutch and entitled *Slang-Brand-Spuiten*. The engraving reproduced depicts the fire at the Amsterdam rope works in 1673. According to van der Heiden the old pumps were used for the last time and the new hoses for the first time at this fire.

4 Newsham's fire engine, 1734

Richard Newsham, of London, in 1721 and 1725 patented 'a new water engine for quenching and extinguishing fires' which was a great improvement on previous machines and was much used in the 18th century. The engine illustrated opposite was built in 1734 and came from Dartmouth. It has two single-acting pumps and an air vessel placed in a tank which forms the frame of the machine.

The pumps were worked by men at the long cross handles, but in addition two treadle boards were provided near the centre of the machine upon which more men stood and assisted the pumping by throwing their weight on to the descending treadle, as illustrated in the background of the frontispiece.

At the front of the engine, protected by a sheet of horn and a door, are directions for keeping the engine in order.

The print reproduced below shows a Newsham engine working at the fire in Cornhill, London, in 1748. The old-style nozzle is in use, but Newsham also supplied sewn leather hose. The use of nozzle and branch connected by hose with a Newsham engine is shown in the frontispiece.

5 London fire engines

During the 18th century and until 1832 fire fighting in London was largely carried out by the fire insurance companies who organised brigades to protect the properties insured by them. The firemen wore splendid and distinctive uniforms and there was great rivalry among the companies to be first at a fire in order to secure the best supplies of water and to advertise the protection which they offered. The insurance brigades were primarily concerned to fight fires threatening the premises of their own policy holders, which were identified by fire marks affixed to the buildings. The brigades did however often help each other and ultimately combined into a single force.

This coloured aquatint, entitled *London Fire Engines. The Noble Protectors of Lives and Property*, after the original painted by James Pollard in about 1825 shows engines of the County, Westminster, and Phoenix companies racing to a fire. The parish engine is to be seen in the background.

> 'The engines thund'red through the street,
> Fire-hook, pipe, bucket, all complete;
> And torches glared, and clattering feet
>   Along the pavement paced . . .
> The Hand-in-Hand the race begun,
> Then came the Phoenix and the Sun,
> The Exchange, where old insurers run,
>   The Eagle, where the new.'

(From *Rejected Addresses* by Horace Smith, 1809.)

CORN DEALER and HAY SALESMAN
HABERDASHER and SILK MERCER
CABINET MAKER UPHOLSTERER & UNDERTAKER
PHŒNIX FIRE OFFICE

6 Horse-drawn manual fire engine, 1866

This is the smallest of the horse-drawn fire engines introduced by Merryweather and Sons in 1851 and known as the *Paxton* after the designer of the Crystal Palace. With 22 men to work it the engine would deliver 100 gallons of water per minute to a height of 120 feet. This engine was purchased in 1866 by the Duke of Portland for his Welbeck Abbey estate. It is exhibited with a figure of a fireman wearing the uniform of the Duke's private brigade.

The pumping machinery and general layout is similar to that illustrated on plate 7, but some modifications in accordance with Shand, Mason & Co. patterns, were made to the carriage after an accident. The firms of Merryweather and Sons, and Shand, Mason, and Co., both descending from 18th century fire-engine makers, were the leading manufacturers in the 19th century and afterwards. The former as a limited company acquired the latter company in 1922.

Similar but larger engines were used by the urban brigades during the nineteenth century. They differed from the parish engines of Newsham's time in several respects. The substitution of metal valves for leather took place in 1792 and by the end of the eighteenth century the use of the treadle had been abandoned. The brigade engines were also mounted on springs, making them suitable for rapid horse traction.

WELBECK.

7 Pumping machinery

The working parts of a manual fire engine are revealed in this sectional model, which was made in 1881 and represents the Merryweather *London Brigade* appliance, a larger version of the type of engine illustrated on plate 6.

The pumping machinery consists of two vertical single-acting pumps driven by links from the horizontal shaft connected with the handles by levers outside the framing. Water is taken in through the inlet valve on the right either through suction hose or through the inlet from the chamber containing the pumps, if this can be kept filled by water from a hydrant. From each cylinder water passes through a delivery valve and out through a branched pipe having a union on each side of the engine. On this pipe there is a large copper air vessel to secure a uniform discharge.

The valves are hinged brass plates, truly ground to fit the circular brass orifices on which they fall. The barrels are of cast brass, with pistons made of two circular pieces of brass, each put into a leather cup and bolted together. The cups are oiled to make a good seal with the cylinder walls.

8 Manual fire engine, 1898

The owners of country houses and the heads of other large establishments were conscious of the lack of protection afforded by the public authorities and many of them set up private fire brigades which flourished during the nineteenth century until the present age of rapid motorised transport. Recognising the importance of prompt action in fire fighting, many firms maintain their own appliances with trained firemen in charge, and there are now more than 500 industrial fire brigades in this country.

Owing to its simplicity and consequent easy maintenance and low cost, the small manual engine was retained for use in country houses, institutions, and small factories, long after the public brigades and larger companies had adopted the steam and later the motor fire engine.

This Merryweather *Factory* engine was supplied to the Office of Works in 1898 for use in Dorchester Prison. It was hand drawn and was worked by twelve men.

9 The first steam fire engine, 1830

The steam fire engine was invented by John Braithwaite, a partner in the engineering firm Braithwaite and Ericsson. Their first engine was built in 1829-1830 and was successfully demonstrated by them at several fires in London. It is depicted in this coloured drawing which bears the signature of the inventor. The steam fire engine met, however, with determined opposition from the conservatively minded fire chiefs, but similar engines were eventually adopted in Liverpool and abroad.

The engine had a boiler and two direct-acting steam pumps and was mounted on wheels for horse traction. The fire box was water-jacketed and was provided with a forced draught by a mechanical bellows and the waste gases issued from the funnel behind the driver's seat. The engine threw 150 gallons of water per minute to a height of 90 ft.

FIRST STEAM FIRE ENGINE CONSTRUCTED IN ENGLAND. A.D. 1830.

BRAITHWAITE AND ERICSSON, INVENTORS AND CONSTRUCTORS, LONDON.

WEIGHT OF ENGINE 2 TONS, 1 QR.

QUANTITY OF WATER THROWN OUT OF A 1¼ INS NOZZLE

1493.3 LBS PER MINUTE, OR 40 TONS PER HOUR TO A HEIGHT OF 90 FEET.

10 Steam fire engine, 1863

This engine, the *Sutherland,* is believed to be the oldest steam fire engine still in existence. It was built by Merryweathers and won the first prize at the international competition held at the Crystal Palace in 1863.

The London Fire Engine Establishment used its first land steam fire engine in 1860 and thereafter interest in such machines developed and competitions were organised in order to discover the best designs. The *Sutherland* proved capable of throwing a powerful jet to over 160 ft. It was purchased by the Admiralty for use at Devonport dockyard. In 1905 it was taken out of service, but was used once again in 1918. The engine has undergone a few alterations since it was built, the chief being the fitting of a newer and different boiler. The photograph below shows the engine in its original state.

1863

11 Steam fire engine, 1894

The *Double Vertical* engine introduced by Shand, Mason & Co., in 1889 succeeded the *London Brigade Vertical*, a single cylinder engine patented by James Shand in 1863. It remained the standard type of horse-drawn fire engine until it was superseded by the petrol motor fire engine.

This engine which came from Southgate, Middlesex, is one of about twenty made to the original pattern and is of the most popular 350 gallon size. It has two double-acting steam cylinders working directly on to two double-acting pumps placed vertically below them. The cranks of the engine are set at right angles so that it starts in any position.

There is an inclined water-tube type of boiler, and steam could be raised from cold water to the working pressure in less than ten minutes while the engine was travelling to a fire.

1894
SOUTHGATE DISTRICT COUNCIL

12 Fire escapes

The model in the centre represents an escape invented by Abraham Wivell, an artist, who used the model in about 1836 when lecturing. The fly ladder was swung into position by means of ropes and access to the main ladder was obtained by opening the hinged section of the fly. A rescued person could be passed down the canvas chute beneath the ladder.

The model on the left, also one of Wivell's, of an extending ladder was less commonly used. Due to his advocacy eighty-five fire escape stations were established in the London district and maintained by the Royal Society for the Protection of Life from Fire.

These ladders were superseded by the lattice-girder extending fire escape which was patented by James Shand in 1880. The model on the right represents a fifty-foot Shand Mason escape manufactured in 1908.

As the fire brigade's prime reponsibility is to save life, an escape is sent to every fire involving a building. In London a pump escape carrying a fifty-foot extending ladder is always included in the first turn-out.

The background scene and models of firemen in uniform typical, from left to right, of 1858, 1836 and 1908 were made recently.

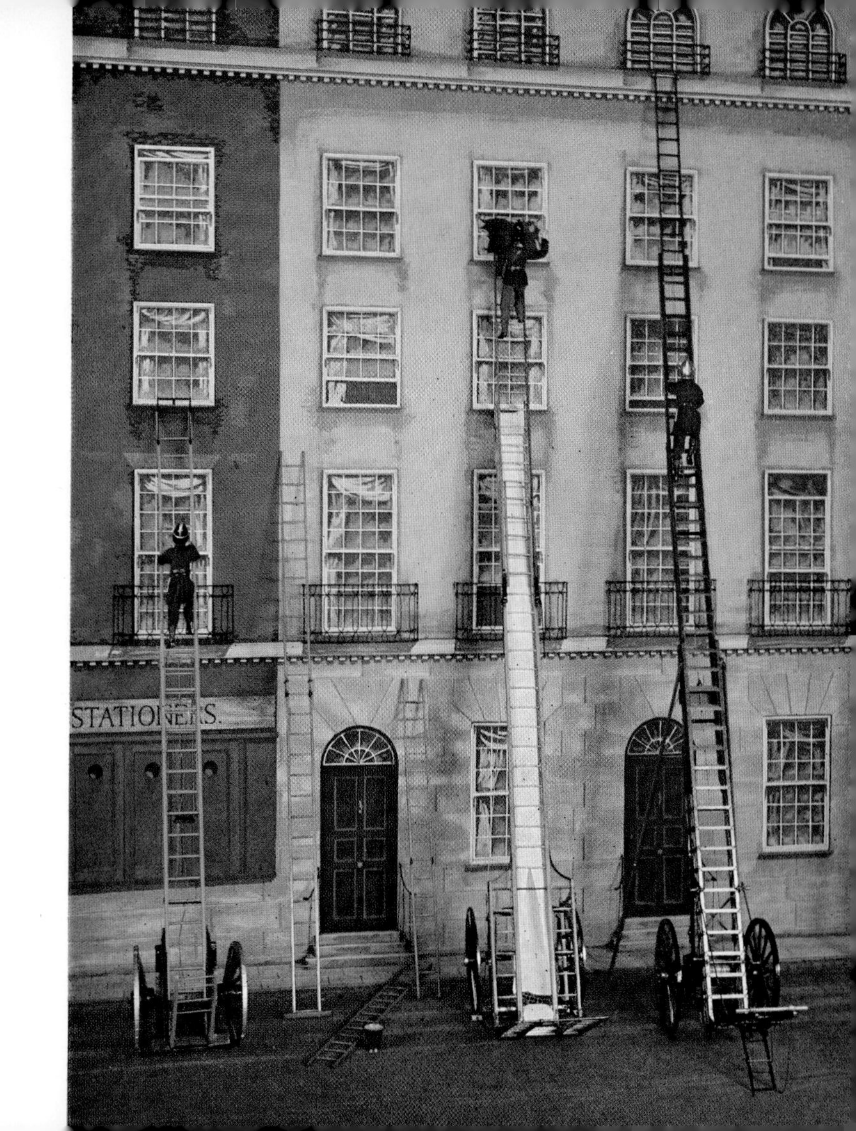
STATIONERS.

13 Chemical engine, 1902

The nineteenth century manual and steam fire engines did not carry a supply of water; so to overcome the delay in getting to work, which occurred while connection was being made to a hydrant, the chemical fire engine was introduced. This was in effect a large horse-drawn soda-acid fire extinguisher and provided a first-aid jet through small-bore tubing.

The model represents the chemical engine used by the Bristol Fire Brigade from 1902 to 1914. It was originally a manual engine made by Merryweather and Sons in about 1888, but was returned to the makers for conversion.

The copper cylinder contained forty gallons of water in which sodium bicarbonate was dissolved. The top fitting contained a lead bottle, filled with sulphuric acid, which was punctured by turning the handle at the rear. The acid reacted with the bicarbonate solution and the carbon dioxide gas generated then expelled the liquid through the tube and nozzle at a pressure of about 100 lb. per sq. in.

BRISTOL

14 Motor fire engine, 1904

This machine was the first self-propelled petrol motor fire engine used by a public fire brigade. It was delivered to the Finchley brigade on 23rd November by Merryweather and Sons, Limited, who had built the very first motor fire engine eight months previously for the Rothschild estate in France.

The power unit was used both for propulsion and for pumping. It was replaced by a more powerful motor in 1912. The *Hatfield* pump has three barrels arranged at 120° to one another, drawing from a common suction chamber and delivering into a common delivery chamber. The tank below the driver's seat is a sixty gallon soda-acid first-aid appliance. Originally a fifty-foot telescopic escape was carried.

This fire engine remained in service until 1928 when it was sold for use as a pump in a gravel pit, whence it was rescued by the Science Museum in 1930. The photograph below was taken in 1904.

15 Motor fire engine, 1936

For many years the standard type of fire-engine body was the *Braidwood*, in which the firemen stood holding on to a rail or sat facing outwards on longitudinal benches. The arrangement had the advantage that the crew could leap off immediately on arrival at a fire. This model represents the Leyland *New World* wagonette fire engine supplied to the Birmingham Fire Brigade. It differs from the earlier fire engines in having the seats facing inwards.
The 700 gallon turbine pump, which is powered by a 115 b.h.p. six cylinder petrol engine, is placed midway between the axles and has suction and delivery branches on both sides. A first-aid apparatus sufficient to extinguish small fires uses a separate gear pump and hose.

16 Limousine type fire engine, 1936

The protection which the enclosed body gives to the firemen so that they arrive at a fire unaffected by a possibly long ride in cold and wet weather is now thought to outweigh the advantages of the open fire engine. The enclosed body was first constructed in 1930 for the Edinburgh Fire Brigade. This model represents a Leyland petrol motor fire engine made in 1936 for the London Fire Brigade.

To increase the pressure from the mains a two-stage centrifugal pump which would deliver 700 gallons per minute of water at 120 lb. per sq. in. was fitted. The pump has two suction and four delivery branches at the rear. The priming gear, comprising two-double-acting piston pumps on the main pump, was used to start a supply from open water.

The full crew consisted of one officer and five firemen.

Many modern fire engines in common with other heavy vehicles are diesel-powered.

Leyland

17 Turntable appliance, 1936

This Leyland-Metz turntable escape, built in 1936, has a ladder which can be extended to 100 feet. The engine drives a 500 gallon per minute turbine pump, which can deliver water to a monitor at the top of the ladder, and supplies power for the training, elevation, extension, and plumbing of the ladder, at the base of which are grouped the controls and indicators. The remote control of the engine throttle is brought about by turning a ring which passes round the fulcrum frame. Levers control the motions of the ladder by means of oil-operated clutches. Over-extension, which might endanger the stability of the ladder, is prevented automatically for any degree of elevation. If the ladder is elevated on uneven ground, relative movement between the ladder and a large plumb weight occurs, which opens a valve and actuates the plumbing mechanism to bring the ladder into the vertical plane.

The turntable ladder (commonly abbreviated to 'T.L.') is employed in a variety of ways: the principal uses being as a water tower and to effect rescues from high buildings, but it can also be used for bridging at very low elevations at right angles to the chassis.

Until 1961 this appliance belonged to the Lancashire Fire Brigade.

DTB 300

18 Portable fire pump, 1966

The lightweight portable pump has been developed to provide at low cost a means of fire fighting which can go into action very quickly before fire brigade appliances could arrive from a fire station at some distance away, and can moreover be used in places difficult of access. A further advantage is that an independent power unit can be made highly efficient and generally more suitable for its work than the ordinary fire engine pump. It is not an entirely satisfactory compromise to use one motor both to propel the vehicle and to drive the pump.

This portable pump made in 1966 by Coventry Climax Engines Ltd., weighs 220 lb. and is therefore light enough to be carried by two men; yet its engine develops thirty horse-power and the centrifugal pump can deliver 250 gallons of water per minute at 100 lb. per sq. in. pressure. This example has been partially sectioned for exhibition.

19 Fire float, 1884

Floating fire engines have been used on the River Thames for the protection of the docks and waterside property since the latter part of the eighteenth century, and indeed a fire engine carried on a boat is illustrated in van der Heiden's book. In 1835 a self-propelled floating steam fire engine was built by John Braithwaite, but it was unacceptable to the London Fire Engine Establishment, and it was not until 1854 that such an appliance was used in London following the successful adaptation of a tug as a fire float at the West India Dock in 1850.

This model represents the *Fire Queen*, which was supplied by Shand, Mason, and Company to the Bristol Corporation in 1884. The vessel was 53 ft. long. The pump had three cylinders directly coupled to three bucket and plunger type water pumps.

Originally two large hose reels were fitted, but these were replaced in 1900 by the monitor. This is a deck-mounted delivery branch which can be directed mechanically as required and can handle high delivery rates at high pressure, where the reaction on a flexible hose would be too great for a fireman to support.

The *Fire Queen* was in service until 1922.

20 Fire boat, 1939

This is a model of the *James Braidwood* which was in the service of the London Fire Brigade from 1939 to 1961. The name commemorates the first Superintendent of the London Fire Engine Establishment who was killed at the Tooley Street fire in 1861.

The boat, which was built by J. Samuel White & Co., of Cowes, is 45 ft. long and has a draught of 3 ft. 6 in. She has three internal combustion engines, each driving a propeller, and is capable of a speed of 20 knots. The two outside engines are also used to drive the turbine fire pumps, each of which has a capacity of 750 gallons per minute at 100 lb. per sq. in. The deliveries run to the monitor and to a swivelling delivery head on each side. The intake is through a strainer on each side of the hull.

There is also a bypass from each delivery to the intake strainer of the other pump, so that water can be forced under pressure to clear the strainer of any blockage.

The old term 'fire float' has generally been replaced by the term 'fire boat'.

Science Museum illustrated booklets

Other titles in this series:

Aeronautica Objets d'Art, Prints, Air Mail

Aeronautics Part 1 : Early Flying up to the Rheims meeting

Aeronautics Part 2 : Flying since 1913

Agriculture Hand tools to Mechanization

Astronomy Globes, Orreries and other Models

Cameras Photographs and Accessories

Carriages to the end of the 19th century

Chemistry Part 1 : Chemical Laboratories and Apparatus to 1850

Chemistry Part 2 : Chemical Laboratories and Apparatus from 1850

Chemistry Part 3 : Atoms, Elements and Molecules

Chemistry Part 4 : Making Chemicals

Fire Engines and other fire-fighting appliances

Lighting Part 1 : Early oil lamps, candles

Lighting Part 2 : Gas, mineral oil, electricity

Lighting Part 3 : Other than in the home

Making Fire Wood friction, tinder boxes, matches

Motor Cars Up to 1930

Physics for Princes The George III Collection

Power to Fly Aircraft Propulsion

Railways Part 1 : To the end of the 19th century

Railways Part 2 : The 20th century

Sewing Machines

Ship Models Part 1 : From earliest times to 1700 AD

Ship Models Part 2: Sailing ships from 1700 AD

Ship Models Part 3 : British Small Craft

Ship Models Part 4 : Foreign Small Craft

Steamships Merchant ships to 1880

Surveying Instruments and Methods

Timekeepers Clocks, watches, sundials, sand-glasses

Warships 1845-1945

Published by
Her Majesty's Stationery Office
and obtainable from the
Government Bookshops listed
on cover page iv (post orders
to PO Box 569, London SE1)

7s each (by post 7s 4d)

Printed in England for
Her Majesty's Stationery Office
by W. Heffer & Sons Ltd
Cambridge

Dd 501696 K80